URANUS, NEPTUNE, and PLUTO

Robin Kerrod

Lerner Publications Company • Minneapolis

This edition published in 2000

Lerner Publications Company
A division of Lerner Publishing Group
241 First Avenue North, Minneapolis MN 55401

Website address: www.lernerbooks.com

Library of Congress Cataloging-in-Publication Data

Kerrod, Robin
    Uranus, Neptune, and Pluto / Robin Kerrod.
        p. cm.
    Includes index.
    Summary: Describes the characteristics of the three most
distant planets in the solar system—Uranus, Neptune, and
Pluto.
    ISBN 0-8225-3908-X (lib. bdg.)
    1. Uranus (Planet)—Juvenile literature. 2. Neptune
(Planet)—Juvenile literature. 3. Pluto (Planet)—Juvenile
literature. [1. Uranus (Planet) 2. Neptune (Plamet) 3. Pluto
(Planet)] I. Title. II. Series: Kerrod, Robin. Planet library.
    QB681.K47   2000   99-39652

Printed in Singapore by Star Standard Industries [PTE] Ltd
Bound in the United States of America
2 3 4 5 6 7 – OS – 07 06 05 04 03 02

CONTENTS

Introducing Uranus, Neptune, and Pluto

Uranus, Neptune, and Pluto are distant and mysterious worlds. All three planets lie many millions of miles away from Earth in the outer reaches of the solar system— the family of bodies that circle in space around the Sun. Of the nine planets in the solar system, Uranus, Neptune, and Pluto are the most distant. In our night sky, Uranus looks like a faint star if viewed with the naked eye. Neptune and Pluto can be spotted only with powerful telescopes.

Scientists have discovered that Uranus and Neptune are giant gas planets, similar to Jupiter and Saturn. Uranus and Neptune are much smaller than the other two gas giants, but they still measure about four times bigger across than Earth. And like Jupiter and Saturn, these planets are made up mainly of gas and liquid, with no solid surface.

The *Voyager 2*
space probe took these
pictures of Uranus
(opposite) and Neptune
(right) in 1986 and 1989.

Astronomers knew very little about Uranus and Neptune until the *Voyager 2* space probe flew past the planets in the 1980s. The probe discovered systems of rings and many moons circling around both planets.

Pluto is quite different from Uranus and Neptune. Compared to its neighbors, Pluto is a tiny planet. In fact, it's the smallest planet in the solar system. Its makeup also differs from that of Uranus and Neptune. Unlike these gas giants, Pluto is mixture of ice and rock. Our most distant planet has no rings and just one moon, named Charon. Remarkably, Charon is half the size of Pluto. Astronomers often call Pluto-Charon a double planet.

Discovering New Worlds

Before the discovery of Neptune in 1781, astronomers believed our solar system contained only six planets.

In ancient times, astronomers watched what they thought were exceptionally bright stars wandering across the night sky. They called these wanderers planets. Ancient astronomers knew of five planets—Mercury, Venus, Mars, Jupiter, and Saturn. These planets could be seen with the naked eye.

In the 1500s, Polish astronomer Nicolaus Copernicus discovered that Earth circled the Sun along with the other planets. The solar system thus consisted of six planets. No one suspected that there were other planets waiting to be discovered.

Above: A drawing in one of Copernicus's notebooks shows Earth and five other planets circling around the Sun.

Right: William Herschel, who discovered Uranus in 1781.

PLANET SEVEN

On March 13, 1781, a young musician and astronomer named William Herschel began his evening stargazing at his home in England. That night he spotted an unusual object in the constellation, or pattern of stars, known as Gemini. At first Herschel thought the object was a comet—an icy body that orbits the Sun and blazes in our sky as it approaches the Sun. But soon Herschel and other astronomers realized that he had discovered a seventh planet. It was eventually named Uranus, after the Greek god of the heavens. The planet proved to be twice as far away from the Sun as Saturn, which was then thought to be the most distant planet. So Herschel's discovery that March night had doubled the size of the solar system.

THE SEARCH CONTINUES

For years astronomers tried to determine the exact path Uranus follows around the Sun, but they found that the planet did not move according to their predictions. They came to the conclusion that another planet must be affecting Uranus's orbit. It was possible, they believed, that the gravity of an unknown planet might be tugging Uranus off course. Gravity is the attraction, or pull, one body has on the objects around it.

Above: In 1845, English mathematician John Couch Adams determined where Neptune could be found.

By the 1840s, two astronomers believed that they had figured out the position of the mystery planet. They were John Couch Adams in England and Urbain Leverrier in France. On September 23, 1846, German astronomer Johann Galle used his telescope to search the part of the sky that Leverrier had suggested. He soon found the mysterious body, planet number eight. It was named Neptune.

THE HUNT FOR PLANET NINE

After Neptune's discovery, astronomers determined that it alone could not account for Uranus's movements. So by the end of the 1800s, the hunt was on for a ninth planet. Among those looking for it was the American astronomer Percival Lowell, who had set up an observatory in Arizona. Lowell had no luck finding the unseen planet before he died in 1916.

Right: Frenchman Urbain Leverrier independently calculated Neptune's location at nearly the same time as Adams did.

Clyde Tombaugh used a 24-inch telescope at the Lowell Observatory (above) in Arizona to photograph the sky. He discovered Pluto by noticing a body that had changed its position among the stars (below).

In 1929, a young astronomer named Clyde Tombaugh began searching for the ninth planet at the Lowell Observatory. His job was to photograph the sky night after night and then examine the photographs to see if any objects had moved among the stars. For many months, Tombaugh took pictures of the region where Lowell had thought the planet might be. He studied these pictures carefully until he found what he was looking for. On February 18, 1930, Tombaugh discovered planet nine, later named Pluto. The observatory announced the discovery of the planet on March 13, 1930—149 years to the day after Herschel had discovered Uranus.

A Lucky Chance?

Astronomers soon realized that Pluto's small size and weak gravity hardly affects Uranus's orbit. It was actually by chance that Lowell's calculations placed Pluto in the constellation Gemini, where Tombaugh later found it. For years afterward, astronomers continued to search for a tenth planet that would explain Uranus's movements. However, more recent studies of Uranus have revealed that its orbit is probably not affected by a tenth planet.

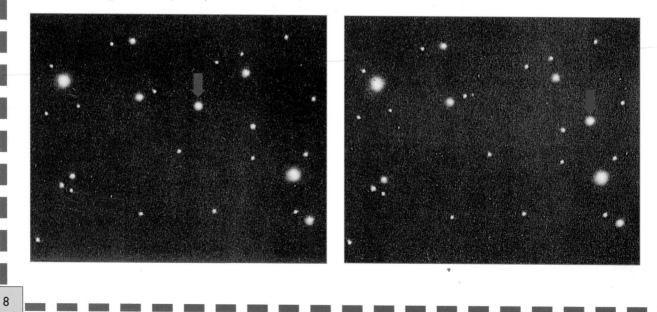

Uranus

Uranus is the third largest planet in the solar system, but its great distance from Earth makes it difficult to see with the naked eye.

Uranus takes about 17 hours to rotate once on its axis.

Uranus is the seventh planet in the solar system, going out from the Sun. At a distance of about 1.8 billion miles (2.9 billion km), it lies twice as far away from the Sun as Saturn. Uranus takes 84 Earth-years to travel once around the Sun.

If you know exactly where to look, you can just barely see Uranus on a clear night. The planet is more easily seen through binoculars or a small telescope. Even in large telescopes, Uranus only appears as a pale greenish body, with no obvious markings.

direction of orbit

direction of rotation on axis

axis

Above: Uranus rotates on an axis that is nearly tilted on its side. Sometimes the planet's poles point directly toward the Sun.

TIPPING OVER

Like the other planets, Uranus rotates, or spins around on its axis. A planet's axis is an imaginary line through it from its north pole to its south pole. But Uranus has an unusual way of rotating on its axis. Most planets rotate on an upright or slightly tilted axis as they orbit.

Uranus's axis is tilted so far over that the planet rotates nearly on its side as it travels around the Sun. This means that the poles on Uranus face either directly toward or away from the Sun. When one pole faces the Sun the other faces away.

Neptune

Uranus

Saturn

Right: Uranus lies in the solar system between Saturn and Neptune. It is twice as far from the Sun as Saturn is and takes nearly three times as long to complete its orbit.

SMALL GIANT

Uranus is the third largest planet in the solar system. With a diameter, or distance across, of about 31,700 miles (51,100 km), it is one of the four giant planets. In fact, Uranus could swallow over 60 bodies the size of Earth. This makes the planet slightly bigger than Neptune but much smaller than the two largest planets, Jupiter and Saturn.

Uranus is about four times bigger across than the Earth. It has more than 15 times the Earth's mass.

INSIDE URANUS

Like Jupiter and Saturn, Uranus is surrounded by a thick atmosphere, or layer of gases. The planet's atmosphere contains mostly hydrogen and helium. A small amount of methane gas in the atmosphere gives Uranus its blue-green appearance.

Astronomers do not know exactly what lies beneath Uranus's atmosphere. But its surface probably differs from that of Jupiter or Saturn. These two giant planets most likely have deep oceans of liquid hydrogen. In contrast, Uranus may have an ocean made up of water, methane, and ammonia. At the center of the planet lies a core of liquid rock about the size of Earth.

Beneath Uranus's thick atmosphere is probably a deep ocean of water and icy gases. The liquid-rock core may contain some metal.

core

liquid layer

atmosphere

THE WEATHER

Compared with Jupiter and Saturn, Uranus has mild weather. Close-up pictures of the planet reveal few weather features in its hazy atmosphere. However, some pictures have shown faint bands of clouds deeper in the atmosphere. These clouds rotate in the

URANUS DATA

Diameter at equator: 31,700 miles (51,100 km)
Average distance from Sun: 1,784,000,000 miles (2,871,000,000 km)
Rotates on axis in: 17 hours, 14 minutes
Orbits Sun in: 84 years
Moons: 17

Magnetic Uranus

All around Earth is an invisible force called magnetism, which is created by our planet's iron core. The magnetism extends into outer space to form a magnetic bubble around Earth known as the magnetosphere. Uranus has a magnetosphere, too. But unlike the magnetosphere on Earth and many other planets, Uranus's magnetosphere does not line up with its north and south poles. Instead, its magnetic poles are closer to the planet's equator. Astronomers have not discovered exactly what produces Uranus's unusual magnetism.

same direction as the planet. They are driven by winds blowing at speeds of up to about 370 miles (600 km) per hour. But pictures of Uranus have not revealed anything like the storms that rage on Saturn and Jupiter.

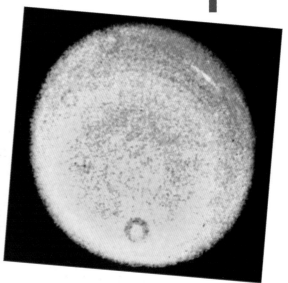

In ordinary photographs (left), Uranus appears to be the same color all over. But a few clouds show up in computer-processed photographs (above).

AN ACCIDENTAL DISCOVERY

On March 10, 1977, American astronomers were flying over the Indian Ocean in a plane carrying a powerful telescope. They planned to observe Uranus as it passed in front of a particular star. By measuring the time Uranus took to pass in front of the star, they could determine the planet's size.

The astronomers switched on their measuring instruments before Uranus was due to pass in front of the star and block out its light. Then a strange thing happened. The star seemed to wink, or dim, several times and then become bright again, as if something were blocking the star's light. After Uranus had passed the star, the star appeared to wink several times again. Astronomers eventually discovered that the winks were caused by a set of rings. Until that time, Saturn was the only planet known to have rings.

An artist's impression of the rings around Uranus. The rings are actually much less bright than they appear here. They are made up of dark material that does not reflect sunlight well.

URANUS'S RINGS

Astronomers have found 11 rings circling around Uranus. The rings begin about 7,000 miles (11,000 km) above Uranus's cloud tops. Altogether, the ring system is about 8,000 miles (13,000 km) wide. The faint innermost ring appears to be made up of fine dust. This ring is about 1,500 miles (2,500 km) wide. The other rings are much narrower, mostly a few miles wide. They are made up mainly of dark boulders that measure a few feet across.

Above: These colorized pictures show slices through Uranus's outermost ring. Computer processing has revealed many ringlets within the ring.

Left: Wide sheets of fine dust appear between Uranus's narrow rings when they are lit from behind by the Sun. The short streaks of light are trails made by distant stars.

Oberon

Ariel

Uranus

Miranda

Umbriel

Titania

URANUS'S MOONS

Scientists have discovered 17 moons orbiting Uranus. Five of them are fairly large and were observed by early astronomers. William Herschel, Uranus's discoverer, discovered the planet's two largest moons in 1787. They were named Titania and Oberon. Two more moons, Ariel and Umbriel, were discovered in 1851, and a fifth moon, Miranda, was spotted in 1948. Uranus's other twelve moons are too small to be seen from Earth. Even the largest, Puck, is only about 95 miles (150 km) across. And the smallest, Cordelia, is only about 15 miles (25 km) across.

Above: The orbits of Uranus's five largest moons. Miranda, the closest of the moons, takes less than a day and a half to orbit Uranus; Oberon takes more than 13 days.

Miranda

Umbriel

Ariel

Titania

The Moon

Oberon

Uranus's five largest Moons are small compared with Earth's Moon.

AMAZING MIRANDA

Miranda is one of the strangest moons in the entire solar system. Its surface is a patchwork of many different kinds of landscape. Some parts of its surface are covered in craters, or deep pits, much like the surface of the Moon. Other parts are rugged, with steep cliffs and deep cracks in the surface. Perhaps the most unusual feature on Miranda is a set of curving grooves that looks like a giant oval racetrack on the moon's surface. Some of the oval grooves measure up to 200 miles (320 km) across.

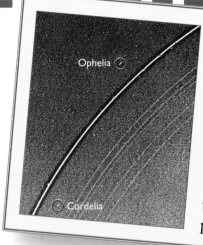

Ophelia

Cordelia

The Little Shepherds

Two of Uranus's tiny moons, Cordelia and Ophelia, orbit close to Epsilon, Uranus's outermost ring. They are called shepherd moons because of the way they seem to herd the ring particles. As these moons travel around either side of the ring, their gravity helps pull back any particles that stray outside Epsilon.

STAR POINT

Most of Uranus's moons are named after characters from the plays of the famous British playwright William Shakespeare.

This is how Uranus might look from Miranda. An artist has added Uranus's rings, although they would probably be too faint to be seen from Miranda.

ARIEL

Ariel is the brightest of Uranus's moons. Sheets of water ice have created bright streaks and patches on its surface. Like the surface of Uranus's other large moons, Ariel's surface is covered with craters. Most craters on this moon are less than 20 miles (30 km) in diameter. Deep cracks and a number of wide, branching valleys with smooth floors also cut into Ariel's surface.

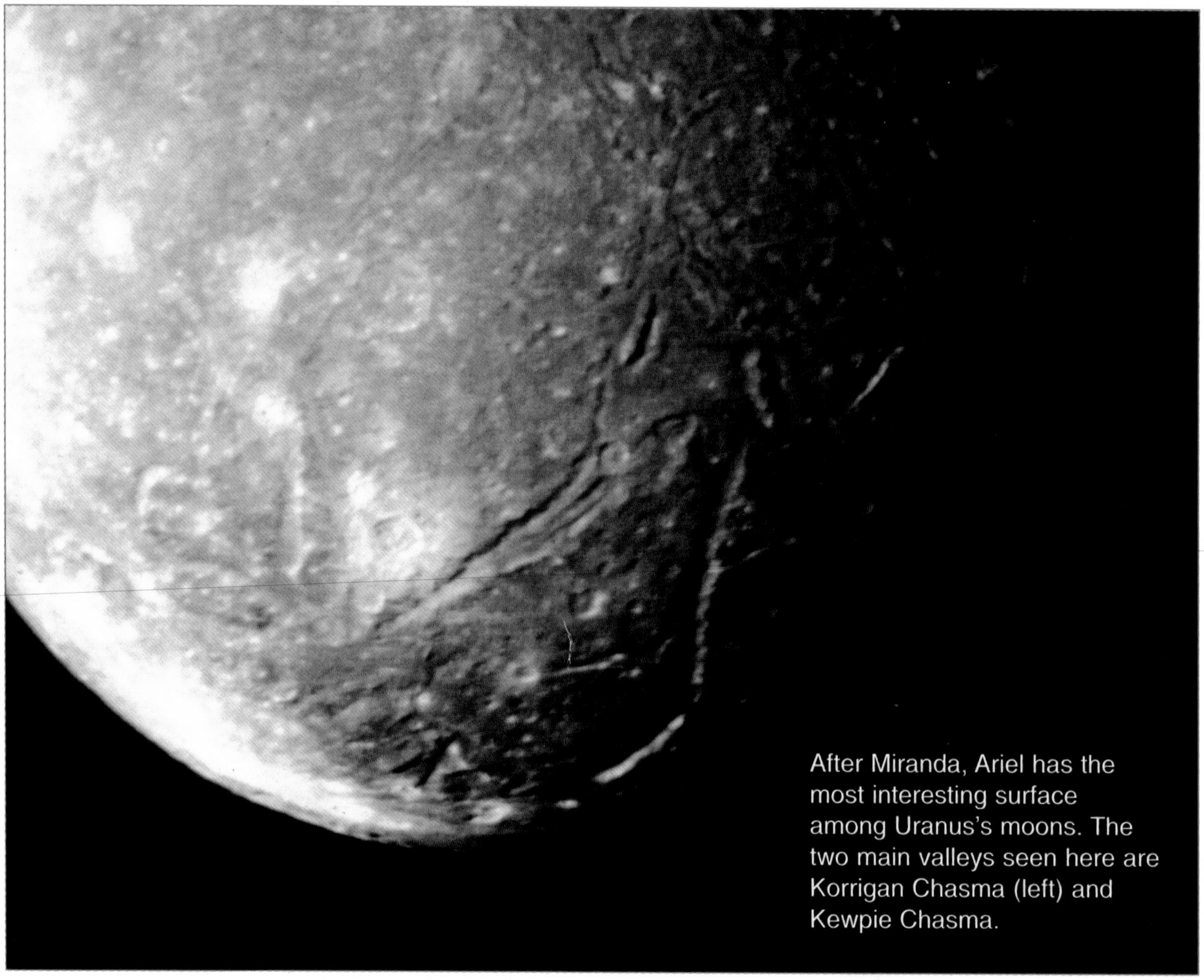

After Miranda, Ariel has the most interesting surface among Uranus's moons. The two main valleys seen here are Korrigan Chasma (left) and Kewpie Chasma.

MOONS OF URANUS DATA

Moon	Diameter (miles)	(km)	Ave. distance from planet (miles)	(km)
Miranda	300	480	80,400	129,400
Ariel	720	1,160	118,700	191,000
Umbriel	730	1,170	165,200	266,000
Titania	980	1,580	271,000	436,300
Oberon	950	1,520	362,300	583,400

Umbriel

Titania

Oberon

UMBRIEL

Umbriel is nearly the same size as Ariel, but it has a much darker surface. In fact, it is the darkest of Uranus's moons. Much of Umbriel's surface is covered with large craters such as Skynd, which is at least 70 miles (110 km) across. The strangest feature on the moon is a bright circle called the "fluorescent Cheerio." Astronomers think that this feature may be the icy rim or floor of a crater.

TITANIA

Titania is Uranus's largest moon. In many ways, it looks similar to Ariel because deep cracks and valleys mark its surface. The largest valley, named Messina Chasma, stretches for more than 900 miles (1,400 km). That's over three times the length of Earth's Grand Canyon in Arizona. Some of Titania's valleys cut through the many craters that cover the moon. The largest craters are up to 185 miles (300 km) wide.

OBERON

Of the large moons, Oberon orbits farthest from Uranus. Deep cracks mark Oberon's heavily cratered surface, and an unidentified dark coating covers the floors of some craters on the moon. The coating may be a mixture of ice and materials containing carbon.

Neptune

The most distant of the gas giants, Neptune is very similar to Uranus in size and makeup.

Neptune is the eighth planet in the solar system. It lies on average about 2.8 billion miles (4.5 billion km) from the Sun and takes nearly 164 Earth-years to complete its orbit. Neptune's day, or the time it takes to rotate once on its axis, is short—just over 16 Earth-hours.

Only slightly smaller than Uranus, Neptune measures about 30,800 miles (49,500 km) across. But this gas giant is too far from Earth to be seen with the naked eye. Even with a powerful telescope, we can see few details on the planet. We have learned what we know about Neptune from the *Voyager 2* space probe that traveled to the outer reaches of the solar system.

Like Uranus, Neptune is about four times bigger across than Earth.

In the solar system, Neptune lies between Uranus and Pluto. At certain times, Pluto journeys inside Neptune's orbit, and Neptune becomes the farthest planet.

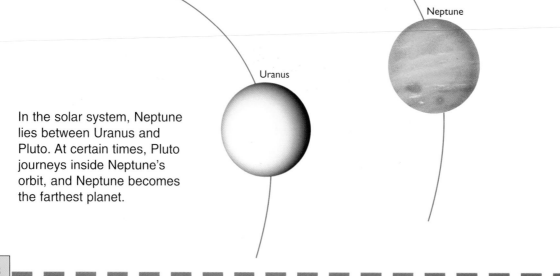

Uranus

Neptune

Pluto

NEPTUNE'S MAKEUP

The blue face of Neptune that we see in photographs is the top of a very thick atmosphere. Similar to the atmosphere of Uranus, Neptune's atmosphere contains mainly hydrogen and helium, with a smaller percentage of methane. As on Uranus, the methane gas gives Neptune its lovely blue color.

A deep ocean of water, methane, and ammonia probably covers Neptune beneath its thick atmosphere. Underneath this ocean lies an Earth-sized core made of liquid rock and ice.

Heated Planet

By our standards, Neptune is a very chilly place. But its temperatures are higher than we would expect for a planet so far from the Sun. That's because Neptune appears to be heated from within. The planet releases more than twice as much heat as it receives from the Sun. At its cloud tops its temperature measures about –350° F (–210° C). This is very similar to temperatures on Uranus, even though Uranus is a billion miles (1.6 billion km) closer to the Sun.

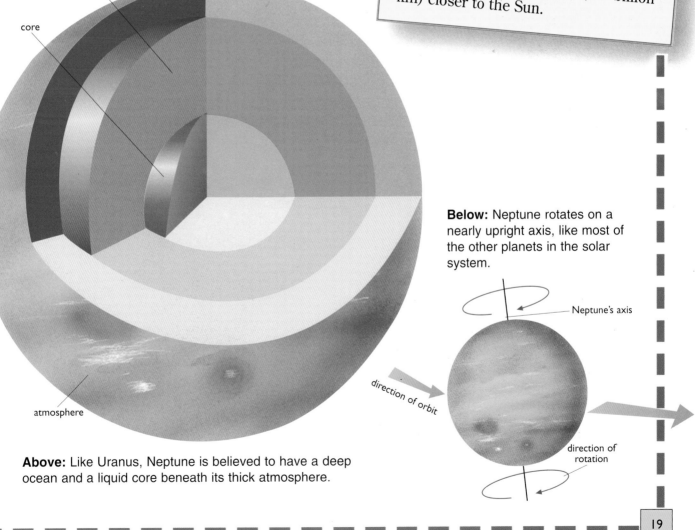

liquid layer

core

atmosphere

Above: Like Uranus, Neptune is believed to have a deep ocean and a liquid core beneath its thick atmosphere.

Below: Neptune rotates on a nearly upright axis, like most of the other planets in the solar system.

Neptune's axis

direction of orbit

direction of rotation

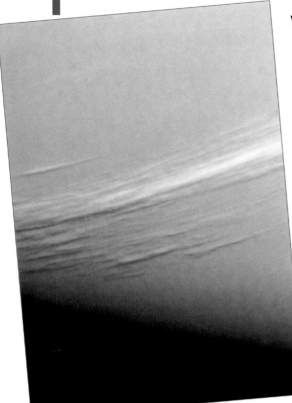

Above: An unusual view of the clouds in Neptune's atmosphere, with the Sun shining on them from a low angle. The shadows they cast make them stand out clearly.

WIND AND CLOUDS

Heat produced within Neptune creates violent winds in its atmosphere. They whip around the planet in the opposite direction of Neptune's rotation. The winds can reach speeds of 1,500 miles (2,400 km) per hour, making them the fastest winds in the solar system.

Along with strong winds, bands of high white clouds rotate in Neptune's atmosphere. The clouds are made up of methane ice, in much the same way that high cirrus clouds on Earth are made up of water ice. One cloud patch, named Scooter, flies around the planet once every 16 hours. It appears to change size and shape over time.

STORMY WEATHER

Furious storms break out in Neptune's windy atmosphere. The storms appear on Neptune as dark oval regions surrounded by wisps of white cloud. In 1989, scientists observed one particularly large storm in the planet's atmosphere. They named it the Great Dark Spot (GDS) after the great storm on Jupiter called the Great Red Spot.

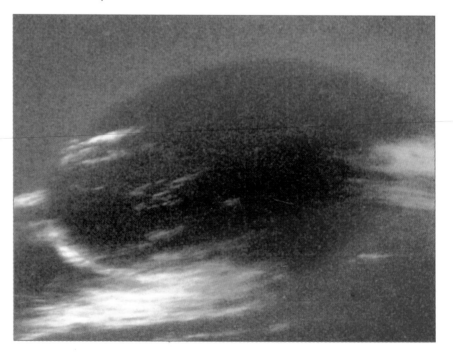

Right: A close-up picture of Neptune's Great Dark Spot, a huge storm system in the planet's atmosphere. It is ringed with clouds.

The GDS was not as big as the Great Red Spot, but it was still very large—about the size of Earth. Winds raced around the GDS at speeds of up to 745 miles (1,200 km) per hour. While Jupiter's Great Red Spot has lasted for hundreds of years, storms on Neptune appear to have short lives. Pictures taken by the Hubble Space Telescope showed that the GDS had disappeared by 1994.

ANOTHER RINGED PLANET?

By the 1980s, scientists had discovered that Saturn, Jupiter, and Uranus all had rings. Astronomers began wondering whether Neptune had rings, too. From Earth, no rings could be seen around the planet.

Astronomers searched for rings around Neptune using the same method that had led them to discover Uranus's rings. They waited for a star to wink just before and just after Neptune passed in front of it. The results of this experiment were confusing. When Neptune passed in front of a star, scientists observed that the star winked sometimes but not all the time. They concluded that instead of having complete rings, Neptune must have arcs, or part-rings.

NEPTUNE DATA

Diameter at equator:
30,800 miles (49,500 km)
Average distance from Sun: 2,794,000,000 miles (4,498,000,000 km)
Rotates on axis in: 16 hours, 7 minutes
Orbits Sun in: 163.7 years
Moons: 8

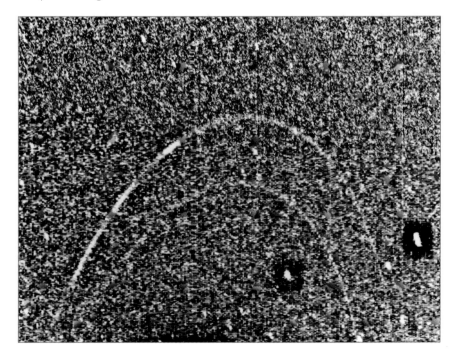

Voyager 2's first clear pictures of Neptune's rings show clumps of bright material that appeared at first to be arcs around the planet.

VOYAGER'S VIEW

When the space probe *Voyager 2* approached Neptune in 1989, it spotted the ring arcs astronomers had detected from Earth. As the probe got closer, astronomers could see that the arcs were actually part of a complete ring. *Voyager 2* discovered that Neptune had four rings in all.

The two main rings are bright and narrow and appear to be made up of fine dust and small particles. The outer ring contains bright clumps of particles. These clumps are what astronomers mistook for the arcs. Neptune's other two rings are wide but much fainter. The one closest to the planet may be at least 1,100 miles (1,700 km) wide. The other faint ring, between the two narrow main rings, measures about 3,600 miles (5,800 km) wide.

Above: This *Voyager* picture shows Neptune's two main rings. The outer ring lies about 39,000 miles (63,000 km) from the planet's center, while the other ring is about 33,000 miles (53,000 km) closer.

NEPTUNE'S MOONS

From telescopes on Earth, only two moons can be seen orbiting Neptune—Triton and Nereid. Triton, Neptune's largest moon, is about two-thirds the size of our Moon. It measures 1,680 miles (2,700 km) across. Nereid is only

Below: Of the four rings that circle Neptune, two are narrow and bright and two are broad and faint.

narrow rings

broad rings

about 210 miles (340 km) across. Neptune's six remaining moons were discovered by the *Voyager 2* space probe. They are Naiad, Thalassa, Despina, Galatea, Larissa, and Proteus.

At 260 miles (420 km) across, Proteus is actually larger than Nereid. It cannot be spotted from Earth because it lies too close to Neptune and gets lost in the planet's glare. The rest of the moons are very small—from 30 to 240 miles (50–380 km) across. These small moons are made up of a dark, rocky material, and they have an irregular shape.

Two views of Neptune's largest moon, Triton. The bottom picture shows the region around the moon's south pole. The inset shows a huge ice-filled crater.

DEEP-FROZEN TRITON

Triton orbits Neptune at a distance of about 220,000 miles (354,000 km) from the planet, about the same distance between the Moon and Earth. However, while the Moon orbits Earth in the same direction as our planet's rotation, Triton orbits in the opposite direction of Neptune's rotation.

Triton is a deep-frozen world with a temperature of only about −390° F (−235° C). This is the coldest known place in the solar system, colder even than distant Pluto. A layer of frozen gases, including nitrogen and methane, covers its icy surface. A pink-colored polar cap rises at the moon's south pole. And icy volcanoes cover parts of the moon where gases, ice, and plumes of dust forced their way to the surface. Material from the plumes settled on the ground and created dark streaks on the surface. Deep cracks, some of them filled with ice, also cut into Triton's frozen landscape.

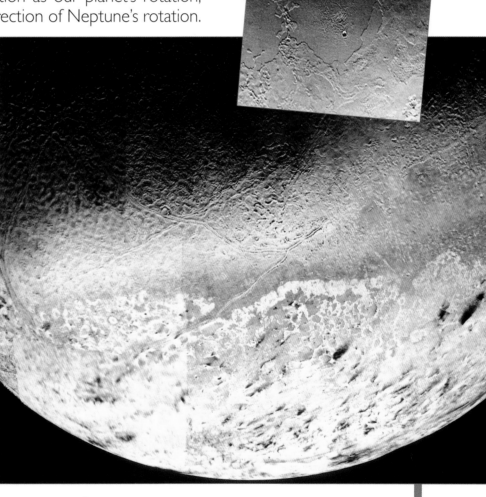

Pluto

Astronomers know very little about tiny Pluto, the most distant of the planets. Some astronomers have wondered if it should be classified as a planet.

Pluto is by far the smallest planet in the solar system. With a distance across of only about 1,480 miles (2,390 km), it's less than half the size of the next largest planet, Mercury, and only two-thirds the size of our Moon.

Most of the time, Pluto is the outermost planet in the solar system. However, its unusual orbit causes it to occasionally change positions with its closest neighbor, Neptune. Pluto orbits the Sun in an extremely elliptical, or oval, orbit—more elliptical than any other planet in the solar system. This means that its distance from the Sun varies greatly, from a near point of about 2.8 billion miles (4.5 billion km) to a far point of about 4.6 billion miles (7.4 billion km). At times, Pluto's oval orbit takes it inside the orbit of Neptune. When this occurs, Neptune becomes the outermost planet.

Pluto's orbit differs from that of the other planets in another way. It travels in a different plane. A plane is like an imaginary sheet of paper in space, with the Sun in the center of the

The Moon

Pluto

Charon

Above: Pluto is smaller than the Moon, and is only twice as big as its own moon, Charon.

Neptune

Pluto

Right and opposite top: Most of the time, Pluto is the farthest planet. But in February 1979, Neptune journeyed outside of Pluto's orbit and became the outermost planet, which lasted until February 1999.

Pluto's orbit

Neptune's orbit

1979 crossing

1980

1999 crossing

2000

paper and the planets orbiting around it. Most of the planets orbit on this same imaginary piece of paper, or the same plane. Pluto orbits in a different plane. Its orbit takes it far above and far below the other planets during its 248-year journey around the Sun.

Some scientists have wondered if Pluto is a planet at all. In the past, astronomers had suggested that Pluto was a moon that escaped from Neptune's orbit. More recently, astronomers have thought that Pluto may be one of the leftover bodies from the solar system's formation. Nearly 5 billion years ago, the solar system formed from a great cloud of gas and dust. Most of this material formed the Sun, the planets, and their moons. The leftover smaller lumps of matter are known as asteroids, comets, and meteors. It is possible that Pluto may be the one of largest of these leftover lumps of matter. However, most scientists have classified Pluto as a planet because of its size, its predictable orbit, and the fact that it has a moon.

LOOKING AT PLUTO

Even in the most powerful telescopes on Earth, distant Pluto looks like a faint star. No details of the planet's surface can be observed from Earth. However, the Hubble Space Telescope, orbiting above Earth's atmosphere, has taken pictures that reveal some surface details. These pictures have shown that Pluto is an icy planet with polar caps. It also has light and dark areas on its surface, probably caused by patterns of ice and frost.

The powerful Hubble Space Telescope is the only instrument that can picture Pluto (left) and its moon, Charon, clearly.

PLUTO'S MAKEUP

Pluto is a dark, cold world made up of a mixture of rock and ice. Temperatures on the planet are around −355° F (−215° C). Frozen methane, nitrogen, and carbon monoxide gas probably make up Pluto's outer layer of ice. Beneath this outer layer lies what may be a layer of water ice. Underneath that may be a rocky core.

Pluto has a very faint atmosphere made of nitrogen, carbon monoxide, and methane—the same gases that make up Pluto's top layer of ice. Astronomers believe that Pluto's atmosphere forms as it approaches the Sun and disappears as it travels away from the Sun. When Pluto is closest to the Sun, the Sun's heat melts some of the ice on Pluto's surface. The melted ice changes into gases that rise from the surface to form the thin atmosphere. As Pluto travels away from the Sun, the gases freeze again and fall back to the surface.

core

frozen methane, nitrogen, and carbon monoxide

frozen water ice

Above: Astronomers think that Pluto is probably made up of several layers of ice surrounding a rocky core.

A DOUBLE PLANET?

In 1978, astronomers studying magnified pictures of Pluto noticed that the planet had an odd shape. Pluto appeared to have a bump on one side. Clearer pictures revealed that the bump was not attached to Pluto. It was actually a moon, which astronomers called Charon.

Remarkably, Charon turned out to be

Left: In telescopes on Earth, the images of Pluto and Charon look blurred together (top left). But the Hubble Space Telescope shows that the two bodies are separate.

half the size of Pluto. No other planet has a moon that is half its size. Until astronomers discovered Charon, our Moon was thought to be the largest moon in comparison with its planet—and it is only one-fourth of Earth's size. Charon is so big compared to Pluto that some astronomers call the two bodies a double planet.

Charon orbits very close to Pluto, only about 12,200 miles (19,600 km) away. It completes an orbit around Pluto in about 6 days and 9 hours. Amazingly, Pluto takes only slightly more time to rotate once on its axis. This means that the planet and moon appear to be locked together as they move through space. Charon remains in nearly the same location above Pluto.

PLUTO DATA

Diameter at equator:
1,484 miles (2,390 km)
Average distance from Sun:
3,700,000,000 miles
(5,900,000,000 km)
Rotates on axis in:
6 days, 9 hours, 36 minutes
Orbits Sun in: 248 years
Moons: 1

An artist has painted this picture of Pluto and Charon. Billions of miles away, the Sun shows up as a bright star.

STAR POINT

If you could visit Pluto, you would only be able to see Charon from one side of the planet. Charon would never be visible from the other side.

Voyager's Distant Mission

Voyager 2 has provided us with most of our information about distant Uranus and Neptune.

In 1977, NASA (National Aeronautics and Space Administration) began its missions to the outer planets with the launch of *Voyager 2* in August and *Voyager 1* in September. Most of the time, the outer planets are too far apart for a space probe to visit all of them in one mission. But the launchings were timed to take advantage of the outer planets' positions in space. *Voyager 1* visited Jupiter and Saturn, while *Voyager 2* was chosen to travel to all four gas giants. Unfortunately, the probe was not able to visit Pluto.

Voyager 2 closes in on Uranus in January 1986. It is heading towards the planet's south pole, which was facing the Sun at the time. It is now so far away that its radio signals, traveling at the speed of light, take 2 hours and 45 minutes to travel back to Earth.

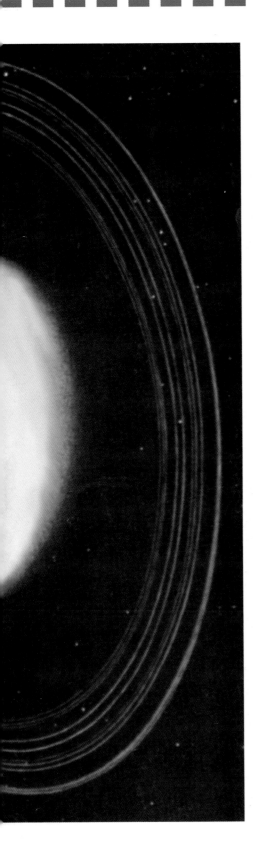

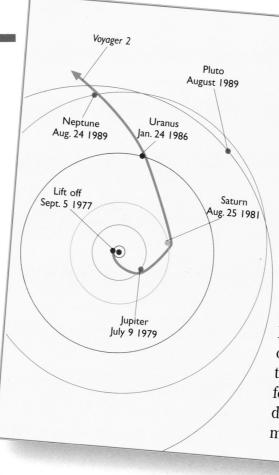

Gravity Assist

The method NASA used to get *Voyager 2* from planet to planet is called gravity assist, or the slingshot method. Scientists planned for the space probe to use the gravity of planets. While passing near a planet, the probe would be speeded up and flung away by the force of the planet's gravity. Scientists were so successful with this method that *Voyager 2* reached its final target, Neptune, within a few seconds of the planned schedule. By then, the probe had traveled for more than 12 years over a distance of more than 4 billion miles (7 billion km)!

THE THIRD ENCOUNTER

In 1981, after visiting Jupiter and Saturn, *Voyager 2* set course for its third target—Uranus. Little was known about Uranus, and astronomers were eager to learn more about this tilted planet.

Over the five years it took *Voyager 2* to travel to Uranus, mission scientists at the Jet Propulsion Laboratory in Pasadena, California, carefully reprogrammed its computer. They wanted to improve communications over the vast distance between Earth and Uranus—about 1.7 billion miles (2.7 billion km).

Voyager 2 began to approach Uranus in 1985 and came closest to the planet in January 1986. The space probe revealed that the planet was an overall blue-green color, with no obvious features. *Voyager 2* also discovered more rings and 10 tiny moons. Its pictures of the moon Miranda caused the greatest surprise. Miranda proved to have one of the strangest surfaces of all the moons in the solar system.

The Sounds of Earth

Do intelligent beings exist in the Universe on planets other than Earth? Scientists don't know the answer to that question, but *Voyager 1* and *Voyager 2* are both carrying a message from Earth just in case. The message takes the form of a record disk, which aliens could play to find out about our planet. It contains music and sounds from nature, coded images of our home planet, and greetings from humans in 55 different languages.

FINAL CALL

After leaving Uranus, *Voyager 2* set course for Neptune. It passed closest to the planet in August 1989, 12 years after it had been launched from Earth. Neptune had its surprises too. It had a deep-blue atmosphere, with clouds and violent storms. *Voyager 2* also discovered a complete set of rings, making Neptune the fourth planet known to have rings.

Voyager 2 discovered several new moons orbiting Neptune—six in all. And its views of Triton revealed that Neptune's largest moon was a deep-frozen world with unusual icy volcanoes.

OFF TO THE STARS

Voyager 2 left Neptune behind in 1989 and began heading out of the solar system. As it travels billions of miles away from Earth, it continues to send back information about conditions in deep space.

On its way to Neptune, *Voyager 2* took this photograph of Uranus.

Glossary

asteroid: a small, rocky body that circles the Sun or another heavenly body

atmosphere: the layer of gases around a planet or another heavenly body

axis: an imaginary line running through a planet from its north to its south pole

comet: a body made up of ice and dust that glows when it approaches the Sun

constellation: a pattern of stars in the sky

core: the center part of a planet or a moon

crater: a pit in the surface of a planet or moon

double planet: a planet and moon system in which the moon is relatively large, as with Pluto and its moon Charon

encounter: a meeting in space, for example between a space probe and a planet.

gas giant: a large planet that is made up mainly of gas; Uranus and Neptune are both gas giants

gravity: the attraction, or pull, that a heavenly body has on objects on or near it

gravity assist: using one planet's gravity to speed up a space probe and direct it to another planet

magnetosphere: a magnetic region in space around a planet

meteor: a small lump of rock or metal from space that produces a streak of light when it enters a planet's atmosphere

NASA: the National Aeronautics and Space Administration, which organizes space activities in the United States

orbit: the path in space of one heavenly body around another, such as Uranus around the Sun

planet: a large body that orbits the Sun

probe: a spacecraft that travels from Earth to explore bodies in the solar system

ring system: a set of rings around a giant planet, made up of rocky and icy particles

shepherd moon: a tiny moon located near a planet's ring that helps keep the ring particles in place

solar system: the Sun and all the bodies that circle around it, including Uranus, Neptune, and Pluto

Index